Manda Education™

Math Boot Camp for the SSAT® Upper

5 Practice Tests & Extremely Difficult Questions

Math Boot Camp for the SSAT® Upper

5 Practice Tests & Extremely Difficult Questions

Editorial:
Justin Grosslight, head author

eBook ISBN: 978-0-9974232-3-5

Note that SSAT is a registered trademark of the Secondary School Admission Test Board, which neither sponsors nor endorses this product.

TABLE OF CONTENTS

WHAT IS THE PURPOSE OF THIS BOOK?

Often students aim to prepare extensively for their SSAT exam, yet the number of available resources for practice is not very readily available. This can prove to be problematic, as the caliber of questions on the SSAT can be extremely challenging – often to the point of being very abstract and well above the grade level of the test-taking student. This book has been developed with these understandings in mind, noting specifically that, though the curricular concepts tested have remained similar, SSAT mathematics questions have become more challenging in recent years. To that effect, the practice tests in this book contain a number of intricate questions that require deep, abstract thought as well as computational mastery. If anything, the harder questions here may even be more challenging than those a student may face on a real SSAT exam. This book is intended for anyone hoping for a good math challenge, with a focus on those students in grades 8 through 10 who plan to take the SSAT upper exam.

A note to consider: our staff at Manda Education™ has years of expertise in test preparation and has constructed some very challenging exams in this book. Our mock exams may be slightly more competitive than other brands of test prep. Parents should not panic about performances in this book or on test day itself that seem out of step with their child's abilities. Though SSAT scores are important, the application process to private American boarding schools is holistic and considers grades, interviews, essays, recommendations, extracurricular activities, and other factors as part of the admissions process. Nonetheless, with proper practice, guidance and encouragement, anybody can improve his or her chances of SSAT success.

Test #1, Section 1

30 Minutes, No Calculator

Following each problem in this section, there are five suggested answers. Work each problem out, look at the five suggested answers, and decide which one is best.

1.   If 5 + 4 − 3 = 6 + 2 − x, then x =

(A)   -2
(B)   -1
(C)   0
(D)   1
(E)   2

2.   Jill needs to buy balloons for her son's birthday. If each of her son's 27 friends and her son each get a balloon and balloons are packaged in boxes of five, what is the minimum number of boxes that Jill must buy?

(A)   5
(B)   6
(C)   7
(D)   8
(E)   9

3.   An angle whose measure is exactly 90 degrees is called:

(A)   a right angle
(B)   a complementary angle
(C)   a straight angle
(D)   an obtuse angle
(E)   an acute angle

4.   What number belongs here: $6 \times 14 \times 3 = 9 \times 4 \times$ ____?

(A)   2
(B)   3
(C)   5
(D)   7
(E)   9

5. Which of the following numbers is divisible by 13?

(A) 739
(B) 741
(C) 743
(D) 744
(E) 747

6. The cost of taking a taxi consists of c cents for the first 2 miles and d cents for each additional quarter mile. In cents, how much does a 5.5 mile trip cost?

(A) $c + 14d$
(B) $14(c + d)$
(C) $3.5(c + d)$
(D) $c + 5.5d$
(E) $c + 3.5d$

7. If N is an even number, which of the following *must* be odd?

I. $N + 1$
II. $(N + 1)^2 - 2N$
III. N^2

(A) I only
(B) II only
(C) III only
(D) I and II
(E) I, II, and III

8. The sum of the ages of 4 people is 61. In five years, what will be the sum of their ages?

(A) 20
(B) 66
(C) 71
(D) 81
(E) 96

4

9. If $n/6$ is a whole number, then n could be:

(A) 125
(B) 126
(C) 136
(D) 146
(E) 155

10. For what value of x will $\sqrt{x+2} = \sqrt{x} + \sqrt{2}$?

(A) $x = -1$
(B) $x = 0$
(C) $x = 1$
(D) $x = 2$
(E) $x = 4$

11. $\frac{5}{6} + \frac{3}{11} + \frac{1}{2} =$

(A) $\frac{9}{11}$

(B) $\frac{20}{33}$

(C) $\frac{15}{19}$

(D) $\frac{53}{33}$

(E) $\frac{9}{33}$

12. The greatest common factor of $6a^2b^3c$ and $9ab^2c^2$ is:

(A) $3ab^2$
(B) $18a^2b^3c^2$
(C) $3a^2b^3c^2$
(D) $9a^2b^3c^2$
(E) $3ab^2c$

13. The average weight of three boxes of bricks is 144 pounds. If two pounds of bricks are added to two of the three boxes, the average weight of the three boxes will become:

(A) 145 pounds
(B) $145\frac{1}{3}$ pounds
(C) 146 pounds
(D) $146\frac{2}{3}$ pounds
(E) 148 pounds

14. How many degrees are there in a regular pentagon?

(A) 180 degrees
(B) 240 degrees
(C) 360 degrees
(D) 540 degrees
(E) 720 degrees

15. If five times a number is less than forty, that number could be all of the following EXCEPT:

(A) 4
(B) 5
(C) 6
(D) 7
(E) 8

16. Let x, y, and z all be positive numbers. If $xy = 2$, $yz = 5$, and $xz = 10$, then $xyz = $ _____?

(A) 10
(B) 17
(C) 20
(D) 50
(E) 100

17. The price of shirt is marked up 10% and then is discounted 10%. The new price of the shirt is:

(A) 10% less than the original price
(B) 1% less than the original price
(C) The same as the original price
(D) 1% more than the original price
(E) 10% more than the original price

18. Which of the following is necessarily true of the numbers A and B if A and B have an average of 70 and B is less than A?

(A) $A + B = 70$
(B) $A - B = 35$
(C) $A - 35 = 35 - B$
(D) $A = B = 35$
(E) none of the above

19. Niall lives 4 miles from the train station. Jacky lives 11 miles from the train station. How far apart do Niall and Jacky live?

(A) 4 miles
(B) 7 miles
(C) 11 miles
(D) 15 miles
(E) It cannot be determined from the information given.

20. Kim has d dogs. Thai has four more dogs than Kim. How many dogs does Thai have?

(A) d
(B) $d - 4$
(C) $4d$
(D) $d + 4$
(E) $4d + 4$

21.   A $4 \times 6 \times 10$ rectangular prism is divided into $2 \times 2 \times 2$ cubes. How many such cubes is the prism divided into?

(A)   20
(B)   25
(C)   30
(D)   60
(E)   120

22.   Armand has 6 pears and Julia has 12. How many pears must Julia give to Armand for them to have the same number of pears?

(A)   2
(B)   3
(C)   4
(D)   5
(E)   6

23.   Which of the following could be the value of N if $\frac{2}{5} + N > 1$?

(A) $\frac{2}{3}$

(B) $\frac{1}{3}$

(C) $\frac{1}{6}$

(D) $\frac{1}{7}$

(E) $\frac{2}{9}$

8

Questions 24 and 25 refer to the bar graph below:

24.   A number of people were surveyed and asked what their favorite type of ice cream was. If they were required to pick one of only the five choices on the graph, how many people were surveyed?

(A)   30
(B)   33
(C)   34
(D)   35
(E)   36

25.   How many more people indicated that their favorite ice cream flavor was vanilla than indicated that their favorite ice cream flavor was mint?

(A)   1
(B)   2
(C)   3
(D)   5
(E)   7

STOP
This is the end of the test.

Test #1, Section 2

30 Minutes, No Calculator

Following each problem in this section, there are five suggested answers. Work each problem out, look at the five suggested answers, and decide which one is best.

1.   If $(A + B) = 6$ and $(B + C) = 10$, then $(A - C) =$ _____?

(A)   -4
(B)   -2
(C)   2
(D)   4
(E)   It cannot be determined from the information given.

2.   If the perimeter of both an equilateral triangle with side x and a regular pentagon with side y is 45, then the ratio $\frac{x}{y}$ is equal to:

(A) $\frac{5}{3}$

(B) $\frac{7}{9}$

(C) $\frac{9}{5}$

(D) $\frac{7}{5}$

(E) $\frac{3}{5}$

3.   The number $79,681 \div 39$ is closest to which of the following numbers:

(A)   2
(B)   20
(C)   200
(D)   2000
(E)   20000

4.   If one-third the weight of a crate is equal to 120 pounds, then what is the weight of three crates of this weight?

(A)   40 pounds
(B)   360 pounds
(C)   540 pounds
(D)   720 pounds
(E)   1080 pounds

5.   The slope of the line $y = \frac{1}{4}x + 3$ is:

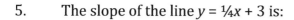

(A)   ¼
(B)   ⅓
(C)   1
(D)   3
(E)   4

6.   If 20% of x is equal to 16, then what is 15% of $4x$?

(A)   32
(B)   48
(C)   64
(D)   80
(E)   320

7.   The average of five consecutive whole numbers is 71. What is the largest of these numbers?

(A)   60
(B)   69
(C)   71
(D)   73
(E)   80

8. A lemonade kiosk has an average of 200 customers per day. In hopes of increasing business, the manager of the kiosk intends to decrease the cost of lemonade from 50¢ to 40¢ after 4:00 P.M. If 120 people purchase lemonade before 4:00 P.M., how many glasses of lemonade must be sold at 40¢ for daily sales to remain the same as before?

(A)   60
(B)   80
(C)   100
(D)   120
(E)   140

9. A store offers a discount of 15% on all shirts. If a shirt initially cost $8.40, what is its discounted price?

(A) $7.03
(B) $7.14
(C) $7.68
(D) $8.00
(E) $8.25

10. Which of the following numbers is the smallest:

(A)   -100
(B)   -3
(C)   -0.001
(D)   0.0001
(E)   47

11. Evaluate: $(2.1 \times 10^3) \times (6.1 \times 10^4) = $ _____?

(A) 8.2×10^7
(B) 8.2×10^8
(C) 1.281×10^7
(D) 1.281×10^8
(E) 1.281×10^9

12. If $16 + S = 17$, then $16 \times S =$ _____?

(A) 15
(B) 16
(C) 17
(D) 18
(E) 19

13. $200 - 18\frac{5}{8} =$

(A) $181\frac{3}{8}$
(B) $181\frac{5}{8}$
(C) $182\frac{3}{8}$
(D) $182\frac{5}{8}$
(E) $182\frac{7}{8}$

14. Round the number 9743.2956801 to the nearest hundredth:

(A) 9743.29
(B) 9743.30
(C) 9743.296
(D) 9700
(E) 9800

15. Express 6% of 9% as a decimal:

(A) 0.09
(B) 0.06
(C) 0.054
(D) 0.0054
(E) 0.54

16. Which number has a reciprocal that is greater than itself?

(A) 1/5
(B) 1/1
(C) 6/5
(D) 9/2
(E) 17/3

17.   What is the greatest number of regions that can be partitioned (eg, made or separated) by three lines in a plane?

(A)   4
(B)   5
(C)   6
(D)   7
(E)   8

18.   What is $5 + 3 \times 2 - 2^3$

(A)   -48
(B)   -8
(C)   -3
(D)   3
(E)   8

19. Which of the following lines is perpendicular to the line $4x + 3y = 12$?

(A) $y = \frac{1}{3}x + 2$
(B) $y = \frac{2}{3}x - 3$
(C) $y = \frac{3}{4}x + 5$
(D) $y = -\frac{3}{4}x + 5$
(E) $y = -\frac{2}{3}x - 1$

20.   Find the product of $3x^3$ and $4x^4$:

(A) $12x^{12}$
(B) $7x^7$
(C) $3x^2 + 4x^4$
(D) $12x^7$
(E) $7x^{12}$

21.   One yard is equal to three feet. How many square yards are in a rectangle that is 15 feet by 18 feet?

(A)   20
(B)   30
(C)   33
(D)   270
(E)   2430

22.   Which of the following statements is true:

(A)   Some squares are rectangles
(B)   All rectangles are squares
(C)   All squares are rectangles
(D)   No rectangles are squares
(E)   No squares are rectangles

23.   What is the smallest integer greater than 1 that is both a perfect square and a perfect cube

(A)   2
(B)   4
(C)   16
(D)   36
(E)   64

24.   If $a \sim\sim b = (2a + b)$, what is the value of $1 \sim\sim (2 \sim\sim 3)$?

(A)   6
(B)   7
(C)   9
(D)   11
(E)   13

25. A chessboard is a square board that consists of a large square divided into 64 small squares (see image below). How many squares of are there in a chessboard?

(A) 64
(B) 65
(C) 91
(D) 140
(E) 204

STOP
This is the end of the test.

Test #2, Section 1

30 Minutes, No Calculator

Following each problem in this section, there are five suggested answers. Work each problem out, look at the five suggested answers, and decide which one is best.

1.   Simplify: $3(x + y) + 5(x - 2y) =$

(A) $3x - 7y$
(B) $8x - 7y$
(C) $5x - 7y$
(D) $3x - 10y$
(E) $7x - 8y$

2.   Which of the following is not true of two lines?

(A)   They can intersect at a point
(B)   They may never intersect
(C)   They can intersect at exactly three points
(D)   They can be the same line
(E)   Each line extends infinitely in both directions

3.   What is the area of a triangle with a base of 9 and a height of 6?

(A)   9
(B)   15
(C)   18
(D)   27
(E)   36

4.   If n > 4, which of the following is the greatest:

(A) $-n$

(B) $\dfrac{1}{n}$

(C) $\dfrac{n}{n+1}$

(D) $\dfrac{n+1}{n}$

(E) $\dfrac{n}{n^2+n}$

5. Erica is three years younger than twice Jake's age. In three years, Jake will be two-thirds of Erica's age. How old was Jake 4 years ago?

(A) 5
(B) 7
(C) 9
(D) 17
(E) 22

6. Evaluate the product $(x + 5)(x - 4)$:

(A) $x^2 - 20$
(B) $x^2 + 9x + 20$
(C) $x^2 + x + 20$
(D) $x^2 + x - 20$
(E) $x^2 - x + 20$

7. Twelve employees at work each contribute equal amounts of money to buy a $24.00 cake for their boss. If four more people decide to pitch in and each person donates an equal amount for the cake, how much less is each person contributing now than before?

(A) $1.00
(B) $0.75
(C) $0.65
(D) $0.50
(E) $0.40

8. $\left(1 - \frac{1}{2}\right) \times \left(1 - \frac{1}{3}\right) \times \left(1 - \frac{1}{4}\right) \times ... \times \left(1 - \frac{1}{100}\right) =$

(A) $\frac{1}{100}$

(B) $\frac{1}{99}$

(C) $\frac{99}{100}$

(D) $\frac{1}{2}$

(E) $\frac{100}{99}$

18

9. At Cedar Mills elementary school, all 240 students must take either Spanish or French class. 164 students take Spanish class and 151 take French class. How many students take both Spanish and French?

(A) 13
(B) 31
(C) 75
(D) 98
(E) 164

10. Below is a grid of city blocks. Annette's home, *A*, is located in the most northwest part of town. Diane's home, *D*, is located in the most southeast part of town. If Annette wants to visit Diane, but can only travel south or east in city blocks (along the grid lines, assuming, of course, that North is up), how many routes are there from Annette's home to Diane's home?

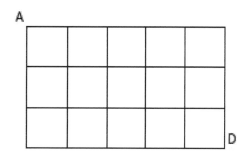

(A) 2
(B) 20
(C) 35
(D) 56
(E) 125

11. Evaluate: $\sqrt{25} + \sqrt[3]{64} + \sqrt[4]{81}$

(A) 1
(B) 12
(C) 16
(D) 17
(E) 21

12. Suppose we have a cube of side length 4. The ratio of the cube's volume to its surface area is:

(A) 1/3
(B) 2/3
(C) 1
(D) 3/2
(E) 3/1

13. Let P be the set of all prime numbers under 60. Let Q be the set of all two-digit numbers whose units and tens digits add to 8 (eg 26, since $2 + 6 = 8$). How many numbers belong to both P and Q?

(A) 0
(B) 1
(C) 2
(D) 3
(E) 4

14. What are the solutions to $(x - 7)(x + 3) = 0$?

(A) $x = 0$
(B) $x = \{0, -3\}$
(C) $x = \{0, 7\}$
(D) $x = 7$
(E) $x = \{-3, 7\}$

15. A stadium is half full. If 2000 more people enter the stadium, the stadium will be 70% full. What is the seating capacity of the stadium?

(A) 2000
(B) 5000
(C) 7000
(D) 9000
(E) 10000

20

16. Which algebraic expression appropriately expresses the phrase "two less than three times a number"?

(A) $2x - 3$
(B) $3x - 2$
(C) $3x + 2$
(D) $2x + 3$
(E) $3x - 2$

17. If the perimeter of a square is 20, what is 140% of the area of the square?

(A) 20
(B) 25
(C) 29
(D) 32
(E) 35

18. A 17-foot ladder is leaning against a building at a point that is 15 feet above the ground. How far from the foot of the building is the foot of the ladder?

(A) 2 feet
(B) 3 feet
(C) 5 feet
(D) 7 feet
(E) 8 feet

19. Suppose that $\frac{1}{7} < x < \frac{1}{3}$. Then the number $\frac{2}{x}$ is between:

(A) 0 and 3
(B) 3 and 6
(C) 6 and 14
(D) 14 and 20
(E) 20 and 25

20. If the ratio of men to women at a party is 3 to 8, then which number of people cannot be at the party:

(A) 132
(B) 140
(C) 143
(D) 154
(E) 165

21. A restaurant owner wants to seat 118 diners at tables of 3 or tables of 4. What is the greatest number of people that can be placed at tables of 4 where all customers are appropriately seated at full tables?

(A) 96
(B) 100
(C) 104
(D) 108
(E) 112

22. At 2:30 P.M., how far apart are the minute and hour hands on a clock?

(A) 120 degrees
(B) 110 degrees
(C) 105 degrees
(D) 100 degrees
(E) 95 degrees

23. Betsy had $44 after spending 11/15 of her pay. How much money did Betsy receive in her paycheck?

(A) $165
(B) $169
(C) $173
(D) $177
(E) $181

24. The average of 9, 11, and 13 is how much more than the average of 2, 6, and 7?

(A) 3
(B) 4
(C) 5
(D) 6
(E) 7

25. If $(m - n)$ is divisible by 5, which of the following must also be divisible by 5?

(A) m
(B) $n/5$
(C) $(m+n)/5$
(D) $5m$
(E) $3n$

STOP
This is the end of the test.

Test #2, Section 2

30 Minutes, No Calculator

Following each problem in this section, there are five suggested answers. Work each problem out, look at the five suggested answers, and decide which one is best.

1. Consider the following sequence of letters: G, N, B, K, G, N, B, K, G, N, B.... If the sequence continues on indefinitely, what will the 703rd letter be?

(A)   G
(B)   N
(C)   B
(D)   K
(E)   None of the above

2. The sum 97 + 231 + 561 + 783 + 46 is how much greater than the sum 95 + 230 + 559 + 782 + 44?

(A)   5
(B)   6
(C)   7
(D)   8
(E)   9

3. How many different, unique 7-letter "words" (or arrangements) can be spelled from the letters RACCOON?

(A)   7
(B)   120
(C)   1260
(D)   2520
(E)   5040

4.   A bag contains 5 red marbles, 6 white marbles, and 4 blue marbles. If I draw
        a marble at random from the bag, what is the probability that it is a blue
        marble?

(A)   4/15
(B)   1/3
(C)   2/5
(D)   5/9
(E)   5/11

5.   A palindrome is a number that is written the same backwards and forwards
        (eg 12321, 90009). How many 5-digit palindromes are there?

(A)   10
(B)   90
(C)   900
(D)   1000
(E)   10000

6.   Which is the correct expansion of $(x - 6)(x + 2)$?

(A) $x^2 - 12$
(B) $x^2 - 8x - 12$
(C) $x^2 - 4x + 12$
(D) $x^2 + 4x - 12$
(E) $x^2 - 4x - 12$

7.   Carla has only dimes and quarters. What is the fewest number of coins that
        she needs to make $2.05?

(A)   9
(B)   10
(C)   14
(D)   17
(E)   19

8.   Which of the following lines is perpendicular to the line $8x + 3y = 10$:

(A) $y = (8/3)x + 1$
(B) $y = (3/8)x - 5$
(C) $y = (-3/8)x + 4$
(D) $y = (5/4)x - 9$
(E) $y = (-8/3)x + 11$

9.   Sara bought a pie and ate 1/5 of it. Her brother Eric ate 2/3 of what was left. After both Sara and Eric had their piece of pie, how much of the original pie remained?

(A)   4/15
(B)   8/15
(C)   2/3
(D)   1/5
(E)   3/5

10.   How many seconds are there in h hours and m minutes?

(A) $60h + m$
(B) $3600h + 60m$
(C) $60h + 3600m$
(D) $h + m$
(E) $60(h + m)$

11.   What is the least common multiple of 5, 6, and 8?

(A)   30
(B)   60
(C)   120
(D)   200
(E)   240

12.   The number of rabbits in a forest doubles every 5 weeks. If there are 100 rabbits in the forest on June 1, how many rabbits are in the forest by September 1?

(A)   Between 200 and 400
(B)   Between 400 and 800
(C)   Between 800 and 1600
(D)   Between 1600 and 3200
(E)   Between 3200 and 6400

13.   An integer between 60 and 92 is a multiple of 6. When divided by 5 it leaves a remainder of 2. When divided by 7, it also leaves a remainder of 2. What is this number?

(A)   66
(B)   72
(C)   78
(D)   84
(E)   90

14.   Luigi is buying hot dog buns, which come in packages of 6. If he needs to feed 57 people each a hot dog, how many packages of hot dog buns does he need?

(A)   10
(B)   11
(C)   12
(D)   13
(E)   14

15.   If $(x + y) = 5$ and $xy = 3$, what is $(x^2 + y^2)$?

(A)   6
(B)   9
(C)   15
(D)   17
(E)   19

16. The price of a chocolate bar rose in the last year from 70 cents to 74 cents. To the nearest percent, what is the price increase of the candy bar?

(A) 4%
(B) 5%
(C) 6%
(D) 7%
(E) 8%

17. In a sequence, each term is 4 greater than the last term. If the first term of the sequence is -95, what is the sum of the tenth and forty-third terms of the sequence?

(A) -144
(B) -55
(C) -38
(D) 14
(E) 23

18. If $0 < A < 1$, which of the following numbers is largest?

(A) $1/A$
(B) $1/(A+1)$
(C) $1/(A-1)$
(D) $1/(2A)$
(E) $-A$

19. A tank with a capacity of 30 gallons is being filled with water at the rate of 1/4 of a gallon every 5 seconds. How long will it take to fill 4/5 of the tank?

(A) 6 minutes
(B) 6.5 minutes
(C) 7 minutes
(D) 7.5 minutes
(E) 8 minutes

20. In cents, what is the value of n nickels, d dimes, and q quarters?

(A) $5n + 10d + 25q$
(B) $0.05n + 0.10d + 0.25q$
(C) $5n + 10q + 25d$
(D) $n + d + q$
(E) $0.01(n + d + q)$

21. There are 12 men and 7 women in an office. How many women would need to be hired in order to make the woman to man ratio in the office 5 to 3?

(A) 9
(B) 10
(C) 11
(D) 12
(E) 13

22. Andy goes on a road trip to the beach. He leaves his home at 9:00AM and arrives at the beach at 1:30PM. He uses 14 gallons of gas. The average number of gallons of gas per hour that Andy uses is closest to:

(A) 2 gallons per hour
(B) 3 gallons per hour
(C) 4 gallons per hour
(D) 5 gallons per hour
(E) 6 gallons per hour

Questions 23 and 24 refer to the following line graph:

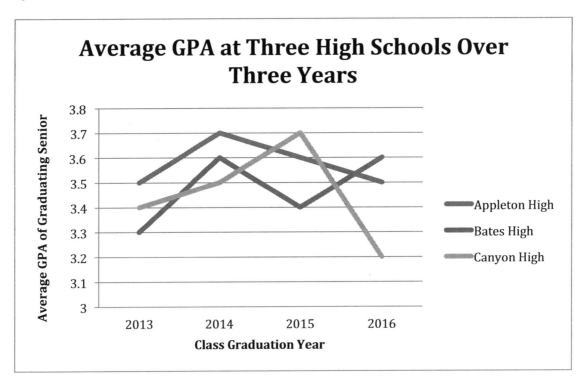

23.   What was the average GPA of a senior graduating from Canyon High in 2015?

(A)   3.3
(B)   3.4
(C)   3.5
(D)   3.6
(E)   3.7

24.   Which school had the highest average GPA of a graduating senior in the class of 2016?

(A)   Appleton High
(B)   Bates High
(C)   Canyon High
(D)   Both Appleton and Canyon High
(E)   It cannot be determined from the information given

25.   The sum of the first 100 positive integers is equal to:

(A)   100
(B)   4950
(C)   5000
(D)   5040
(E)   5050

STOP
This is the end of the test.

Test #3, Section 1

30 Minutes, No Calculator

Following each problem in this section, there are five suggested answers. Work each problem out, look at the five suggested answers, and decide which one is best.

1. If someone randomly picks a number from the set of integers from 40 to 79 inclusive, what is the probability that the number is prime?

    (A)   0.225
    (B)   0.25
    (C)   0.275
    (D)   0.3
    (E)   0.325

2. If $0.2 + N > 1$, which value of N would make the statement true?

    (A)   11/13
    (B)   8/11
    (C)   4/5
    (D)   2/3
    (E)   3/4

3. Peter decides to go for a run after school. He runs 4 miles north, then 7 miles east, then 6 miles south, and finally one more mile east. If a straight-line path were to be drawn, how far is peter from the school?

    (A)   10 miles
    (B)   18 miles
    (C) $\sqrt{53}$ miles
    (D) $\sqrt{68}$ miles
    (E) $\sqrt{149}$ miles

4.   If it takes 3 cups of chocolate chips to make 40 cookies, how many cups of chocolate chips will I need to make 140 cookies?

(A)   8
(B)   9
(C)   9.5
(D)   10
(E)   10.5

5.   A line has an x-intercept of -3 and a y-intercept of 5.   Consider the following statements:

   I.   the line has a positive slope
   II.   the line passes through quadrant IV
   III.   the line passes through (2, 3)

   Which of the following is true:

(A)   I only
(B)   II only
(C)   III only
(D) I and II
(E)   I and III

6.   There are 21 students in a classroom. What could be the ratio of girls to boys?

(A)   3 to 2
(B)   4 to 5
(C)   11 to 1
(D)   4 to 3
(E)   2 to 7

7.   The lengths of the sides of a pentagon are 8, x, $(x+3)$, $(2x – 1)$, and 6 centimeters. If the perimeter of the pentagon is 70 centimeters, what is the value of x?

(A)   13
(B)   13.5
(C)   14
(D)   14.5
(E)   15

8.   What is the least common multiple of $3a^2b^5$ and $5ab^3c$?

(A) $15a^2b^5c$
(B) $15ab^3$
(C) $15ab^2c$
(D) $15ab^5c$
(E) $15a^2b^3c$

9.   What is the greatest whole number less than 70000 that contains all five distinct even digits?

(A)   62840
(B)   84260
(C)   68240
(D)   68420
(E)   86420

10.   A rectangle has an area of 100 and its length is four times its width. What is the length of the rectangle?

(A)   5
(B)   10
(C)   15
(D)   20
(E)   25

11.   A shirt went on sale at 20% off and then was marked up by 30%. What is the new cost of the shirt relative to its original price?

(A)   10% more than the original price
(B)   4% less than the original price
(C)   The same as the original price
(D)   2% more than the original price
(E)   4% more than the original price

12. Each of 12 teams in a soccer league plays every other team three times. How many total games are played?

(A) 33
(B) 66
(C) 132
(D) 198
(E) 396

13. Robin has 11 cupcakes and Scott has 19 cupcakes. How many cupcakes must Scott give Robin in order for them to have the same number of cupcakes?

(A) 2
(B) 3
(C) 4
(D) 5
(E) 6

14. Simplify: $2(x + y) + 3(x + 1) - y$

(A) $5x + y + 3$
(B) $5x + 2y + 3$
(C) $3x + y + 3$
(D) $5x + 2y + 1$
(E) $5x + y + 1$

15. What is the sum of the first 15 positive odd numbers?

(A) 120
(B) 180
(C) 225
(D) 250
(E) 300

16.   A truck traveling at 55 miles per hour can travel how far in 18 minutes?

(A)   15 miles
(B)   16.5 miles
(C)   18 miles
(D) $18\frac{1}{3}$ miles
(E)   20 miles

17.   Let S be the set of all prime numbers less than 20 and T be the set of all positive multiples of three less than 30. If I form a fraction s/t, where s is an element in S and t is an element in T, what is the smallest fraction (in lowest terms) that I can form?

(A)   2/27
(B)   3/30
(C)   1/9
(D)   1/15
(E)   5/29

18.   Martha bought a rectangular piece of land that was 200 feet wide and 300 feet long for $20000. To the nearest cent, how much did she pay per square foot?

(A) $0.30 per square foot
(B) $0.33 per square foot
(C) $0.37 per square foot
(D) $0.41 per square foot
(E) $0.45 per square foot

19.   If $(a - b) = 4$, $(b + c) = 13$, and $(c + d) = -2$, then what is $(a - d)$?

(A)   19
(B)   17
(C)   15
(D)   13
(E)   It cannot be determined by the information above.

36

20.   Tom buys a 200-page book. Each page is numbered beginning at 1 and ending at 200. How many digits are used to write all of the page numbers combined?

(A)   480
(B)   483
(C)   489
(D)   492
(E)   495

21.   A rectangle has a length of 8 and a width of 5. If the length is decreased by 10% and the width is increased by 20%, how does the area of the new rectangle compare with the area of the original rectangle?

(A)   The new rectangle has 8% smaller area
(B)   The new rectangle has 4% smaller area
(C)   Both rectangles have the same area
(D)   The new rectangle has 4% larger area
(E)   The new rectangle has 8% larger area

22.   Consider the following infinite sequence: 2, 5, 8, 11, 14, 17.... What is the 303rd number in the sequence?

(A)   902
(B)   905
(C)   908
(D)   911
(E)   914

23.   Teena is standing in line for ice cream. She realizes that she's the ninth person from the front of the line and also the ninth person from the back of the line. How many people are waiting in line for ice cream?

(A)   17
(B)   18
(C)   19
(D)   20
(E)   21

24. Erica makes $15 per hour working for 40 hours each week and 1.5 times her usual rate for each additional hour. If Erica received a paycheck for $802.50 for last week's work, how many hours did she work last week?

(A) 41
(B) 43
(C) 47
(D) 49
(E) 51

25. Two hoses are filling a swimming pool. If one hose takes 4 hours to fill the pool and another hose takes 6 hours to fill the pool, how long will it take both hoses to fill the pool running at the same time?

(A) 1 hour 54 minutes
(B) 2 hours 24 minutes
(C) 3 hours
(D) 3 hours 30 minutes
(E) 3 hours 50 minutes

STOP
This is the end of the test.

Test #3, Section 2

30 Minutes, No Calculator

Following each problem in this section, there are five suggested answers. Work each problem out, look at the five suggested answers, and decide which one is best.

1.   Lucy goes to the market and buys a carton of milk for $2.79, a package of noodles for $2.19, a steak for $4.74, and a box of cookies for $3.18. How much did she spend in all?

(A) $12.90
(B) $13.50
(C) $14.18
(D) $14.46
(E) $15.04

2.   If the number $(A + B)$ is odd, which of the following must be true?

(A) A is odd
(B) B is odd
(C) $(A + B)^2$ is odd
(D) $(A - 3)$ is odd
(E) $5B$ is odd

3.   The circumference of a wheel is 60 inches. If 12 inches are in a foot and a mile is equal to 5,280 feet, how many revolutions does the wheel make in traveling a mile?

(A)   1014
(B)   1022
(C)   1040
(D)   1056
(E)   1068

4. Alice and Katy are standing at a flagpole. At 3:00PM, Alice begins walking directly east of the flagpole at 4 miles per hour. At the same time, Katy starts walking north at three miles per hour. At 5:00PM, what will be the straight-line distance between the two ladies?

(A) 5 miles
(B) 7 miles
(C) 10 miles
(D) 12 miles
(E) 15 miles

5. Evaluate $(-1)^{101} - (-1)^{102} + (-1)^{103} - (-1)^{104} - (-1)^{105}$:

(A) -3
(B) -2
(C) -1
(D) 0
(E) 1

6. Sandeep would like to order a pizza. He can choose one of two possible crusts, one of three types of sauce, and one of four types of cheese. How many different possibilities for a pizza are there?

(A) 4
(B) 6
(C) 8
(D) 12
(E) 24

7. If $1/a + 1/b = 2/(a + b)$, which of the following must be true:

(A) a and b must be equal
(B) $a = -b$
(C) $a^2 = 0$
(D) $b^2 = 0$
(E) This can never be true.

8.	A diagonal of a regular polygon is any line from a vertex of that polygon to another vertex that does not form one of the sides of the polygon. How many diagonals does an octagon have?

(A)	14
(B)	16
(C)	18
(D)	20
(E)	22

9.	Suppose that $-1 < B < 1$ and consider the following statements:

	I.	$1/B$ can be greater than 94
	II.	$1/B$ can be undefined
	III.	$1/B$ can equal -1

Which of the following statements is/are true?

(A)	I only
(B)	II only
(C)	I and II
(D)	II and III
(E)	I, II, and III

10.	In a rectangle, 1/5 of the area is 2/3 square units. How many square units is 5/12 the area of the rectangle?

(A)	5/12 units
(B)	11/18 units
(C)	18/25 units
(D)	25/18 units
(E)	18/11 units

11. Michael has received grades of 91, 88, 87, and 86 on his four exams. His fifth and final exam (scored in only whole number points) is worth twice the weight of an individual exam. What is the minimum score he needs to receive on the final exam to receive a class grade of at least 90?

(A) 90
(B) 92
(C) 94
(D) 96
(E) 98

12. Solve $5x + 6 = 931$:

(A) $x = 182$
(B) $x = 183$
(C) $x = 184$
(D) $x = 185$
(E) $x = 186$

13. Let x and y be positive integers. How many solutions (x, y) are there to $2x + 3y \leq 19$

(A) 20
(B) 21
(C) 22
(D) 23
(E) 24

14. Marcus goes to a book sale where all books are the same price. The sales clerk tells him that 5 books cost 3.50. How much should 23 books cost?

(A) $5.00
(B) $16.10
(C) $20.45
(D) $23.00
(E) $24.60

15. Define the operation $(x \sim\sim y) = xy - x$. If $(x \sim\sim y) = 0$, then which of the following <u>must</u> be true?

    I. x and y both equal 0
    II.   either x or y must be 0
    III.   neither x nor y is 0

(A)   II only
(B)   III only
(C)   I and II
(D)   I, II, and III
(E)   none of the above are necessary

16. If $2(x + 3) + y + 7x = 2y - 1$, what is y in terms of x:

(A) $y = 9x + 7$
(B) $y = -9x + 7$
(C) $y = 9x + 4$
(D) $y = -9x - 4$
(E) $y = 7x + 9$

17. What is the measure of each angle of an equilateral triangle?

(A)   30°
(B)   45°
(C)   60°
(D)   90°
(E)   120°

18. If 5 workers can build 5 pools in 5 days, how many pools can 3 workers build in 20 days?

(A)   5 pools
(B)   8 pools
(C)   12 pools
(D)   14 pools
(E)   15 pools

19. Fiona, Gianna, and Harold are children. When combined, Fiona and Gianna weigh 39 pounds, Gianna and Harold weigh 42 pounds, and Fiona and Harold weigh 45 pounds. How much does Fiona weigh?

(A) 24 pounds
(B) 21 pounds
(C) 18 pounds
(D) 15 pounds
(E) 12 pounds

20. If $2/11 + 3/N = 4/7$, then $N =$

(A) 77/10
(B) 10/77
(C) 8/77
(D) 77/8
(E) 10/11

21. If four fair coins are tossed at random, what is the probability of getting exactly 2 heads and 2 tails?

(A) 1/4
(B) 3/8
(C) 1/2
(D) 9/16
(E) 5/8

22. The product 0.31×242 is closest to:

(A) 0.8
(B) 8
(C) 80
(D) 800
(E) 8000

23. If $2x + y = 6$, what is $-3x - (3/2)y + 14$?

(A) -9
(B) -5
(C) 4
(D) 5
(E) 9

24. Let *ABCD* be a square with side length 8. *E*, *F*, *G*, and *H* are the respective midpoints of the sides of the square. What is the area of *EFGH*?

(A) 16
(B) 32
(C) 64
(D) 96
(E) 128

25. If a dinner that costs \$200 is to be shared evenly between a party of guests, then which of the following cannot be the number of guests at the dinner?

(A) 10
(B) 12
(C) 16
(D) 20
(E) 25

STOP
This is the end of the test.

Test #4, Section 1

30 Minutes, No Calculator

Following each problem in this section, there are five suggested answers. Work each problem out, look at the five suggested answers, and decide which one is best.

1.   Which of the following is true about any two consecutive integers?

(A)   Their sum is always odd
(B)   Their sum is always even
(C)   Their sum is always prime
(D)   Their product is always odd
(E)   Their quotient is always less than 1

2.   What is the units digit of 2^{2016}?

(A)   2
(B)   4
(C)   6
(D)   8
(E)   0

3.   In how many different possible orderings can 5 people be seated at a round table?

(A)   5
(B)   10
(C)   24
(D)   30
(E)   60

4.   Which of the following numbers is smallest?

(A)   0.001
(B)   0.0001
(C)   0.00001
(D)   -0.001
(E)   -0.0001

5.   The area of a rectangle with length 9 units and width 6 units is the same a
one quarter the area of a square having a side length of how many units?

(A)   6
(B)   8
(C)   10
(D)   12
(E)   14

6.   Which value of N will make the statement true: $1/5 + N > 1$

(A) $N = 3/7$
(B) $N = 15/19$
(C) $N = 16/17$
(D) $N = 10/13$
(E) $N = 4/9$

7.   A parking lot is filled with cars and bicycles. If there are 30 vehicles in the lot
having a total of 102 wheels, how many cars are in the lot?

(A)   19
(B)   20
(C)   21
(D)   22
(E)   23

8.   What is the area in square units of the triangle formed by the x-axis, the y-
axis and the line $5x - 4y = 20$?

(A)   5
(B)   10
(C)   15
(D)   20
(E)   25

9.   What is the greatest common factor of 18, 66, and 240

(A)   2
(B)   3
(C)   6
(D)   9
(E)   12

10.   To enter my high school at night, I need to enter a 3-digit code, where each digit can be from 0-9. No digit can be repeated. What is the difference between the largest possible code I can create and the smallest possible code that I can create?

(A)   986
(B)   975
(C)   979
(D)   990
(E)   989

11.   Evan has taken 5 tests for his math class. He has scored 79, 93, 88, 84, and 87 on them. Assuming that the maximum score Evan can get on a test is 100, that scores are given a whole number score, and that an "A" is a grade of 90 or more, what is the minimum number of additional tests that Evan needs his teacher to assign to ensure any hope of getting an "A"?

(A)   0
(B)   1
(C)   2
(D)   3
(E)   4

12.   If I put $100.00 in a bank account that gains 7% compound interest annually, how much money to the nearest cent can I expect to have after three years?

(A) $107.00
(B) $121.00
(C) $122.00
(D) $122.50
(E) $122.75

13.   Evaluate the following: $(x + 4)^2$:

(A) $x^2 + 8x - 16$
(B) $x^2 - 8x - 16$
(C) $x^2 + 8x + 16$
(D) $x^2 - 16$
(E) $x^2 + 16$

14. A photocopy shop charges 13 cents per copy for the first 500 copies, 9 cents per copy for the next 400 copies, and 7 cents per copy for every copy thereafter. How much should it cost to make 1425 copies?

(A) $121.25
(B) $129.75
(C) $137.75
(D) $148.25
(E) $185.25

15. Let S be the set {1, 2, 3, 5, 6, 7, 10, 12, 13, 15, 17, 18, 19}. What is the probability that if I draw a number from S, it will be a 2-digit prime number?

(A) 7/13
(B) 2/3
(C) 3/7
(D) 2/13
(E) 3/13

16. If I roll three fair dice, what is the probability that the sum of the numbers on their faces equals 4?

(A) 1/216
(B) 1/72
(C) 1/36
(D) 1/9
(E) 1/6

17. Darcy takes a 4.5-hour flight from Chicago to Los Angeles, which is two hours earlier than Chicago. If Darcy's plane departs Chicago at 1:13 P.M. local time, at what time local time will her plane land in Los Angeles?

(A) 5:43 P.M.
(B) 4:43 P.M.
(C) 3:43 P.M.
(D) 2:43 P.M.
(E) 11:43 A.M.

18. Solve for x: $\frac{3}{11} \div \frac{x}{5} = 2\frac{7}{9}$

(A) 26/55
(B) 27/55
(C) 4/11
(D) 55/27
(E) 55/26

19. Which of the following fractions is equals to 0.2791791791791… (Where the decimal digits 791 repeat indefinitely)?

(A) 791/990
(B) 2791/9999
(C) 2791/9990
(D) 2789/9990
(E) 2790/9990

20. The coordinates (0, 1), (1, 0), (0, -1), and (-1, 0) form vertices of a square. What is the area of this square?

(A) 1 square unit
(B) $\sqrt{2}$ square units
(C) 2 square units
(D) $2\sqrt{2}$ square units
(E) 4 square units

21. There are 6 black marbles, 2 white marbles, and 3 yellow marbles in a jar. What is the probability that I draw a black marble and then a white marble (in that order), assuming that no marble is replaced after the first draw?

(A) 12/121
(B) 6/55
(C) 8/121
(D) 4/55
(E) 12/55

22. Jason has 16 socks in a drawer. Each sock belongs to one of 8 different colored pairs. If Jason blindfolds himself and begins picking socks out of the drawer at random, what is the minimum number of socks that he must draw to ensure that he has a matching pair?

(A)   3
(B)   8
(C)   9
(D)   10
(E)   15

23. If March 7 falls on a Monday, on what day will December 6 of the same year fall?

(A)   Tuesday
(B)   Wednesday
(C)   Friday
(D)   Saturday
(E)   Sunday

24. Two ships leave a port at the same time. One ship sails in the northeast direction at 20 miles per hour. The other ship sails in the southeast direction at 20 miles per hour. After 24 minutes, how far apart will the ships be?

(A)   8 miles
(B) $8\sqrt{2}$ miles
(C) $8\sqrt{3}$ miles
(D)   16 miles
(E) $16\sqrt{2}$ miles

25. Consider the set of equations below, where *a*, *b*, and *c* are positive integers:

-*ax* + *by* = *c*
ax + *by* = *c*

Which of the following is true about the graphs of these two lines:

(A)   they will be parallel
(B)   they will be the same line
(C)   they will intersect and be perpendicular
(D)   they will intersect but not be perpendicular
(E)   they cannot be graphed

STOP
This is the end of the test.

Test #4, Section 2

30 Minutes, No Calculator

Following each problem in this section, there are five suggested answers. Work each problem out, look at the five suggested answers, and decide which one is best.

1.   Two numbers are said to be relatively prime if they share no common prime factors. Which two numbers are relatively prime?

(A)   4 and 36
(B)   14 and 33`
(C)   15 and 65
(D)   26 and 78
(E)   22 and 48

2.   Which of the following fractions is equal to 0.54

(A)   27/50
(B)   13/25
(C)   14/25
(D)   17/25
(E)   29/50

3.   If I take the length of the side of a square and triple it, by what percent does the area of the square increase?

(A)   100%
(B)   300%
(C)   600%
(D)   800%
(E)   900%

4. Jamal likes to play basketball and is excellent at shooting free throw shots. The probability that he makes a free throw shot is 4/5. If given the chance to make three free throw shots, what is the probability that Jamal makes exactly two of them?

(A) 16/125
(B) 48/125
(C) 64/125
(D) 2/3
(E) 4/5

5. $45.829 =$

(A) $40 \times 10 + 5 \times 1 + \frac{8}{10} + \frac{2}{100} + \frac{9}{1000}$

(B) $40 \times 1 + 5 \times 1 + \frac{8}{100} + \frac{2}{1000} + \frac{9}{10000}$

(C) $4 \times 10 + 5 \times 1 + \frac{82}{10} + \frac{9}{1000}$

(D) $4 \times 10 + 5 \times 1 + \frac{8}{10} + \frac{2}{100} + \frac{9}{1000}$

(E) $4 \times 10 + 5 \times 1 + \frac{2}{10} + \frac{8}{100} + \frac{9}{1000}$

6. Suzanna goes to the mall once every three days. Jodi goes to the mall once every 5 days. If Suzanna and Jodi are both at the mall on January 10, when will be the next day that they're both at the mall?

(A) January 13
(B) January 15
(C) January 19
(D) January 22
(E) January 25

7. Consider the 6-digit number 37A582. What are all the values for A that will make this number divisible by 3?

(A) 2 or 5
(B) 1, 4, or 7
(C) 5 or 8
(D) 2, 5 or 8
(E) 3 or 9

8.   How many different ways can Jeanette choose to invite exactly three of her six friends to her birthday party?

(A)   3
(B)   6
(C)   20
(D)   60
(E)   120

9.   Suppose that all books have a fixed price and all pencils have a different fixed price. If two pencils and three books cost $12.39, how much will three pencils and four books cost?

(A) $13.70
(B) $14.51
(C) $16.43
(D) $17.89
(E)   It cannot be determined from the information given.

10.   The formula for converting temperatures in Celsius, C, to temperatures in Fahrenheit, F, is given by the following equation: $\frac{9}{5}C + 32 = F$. At what temperature will a Celsius and a Fahrenheit thermometer read the same value?

(A)   100 degrees Celsius
(B)   65 degrees Celsius
(C)   20 degrees Celsius
(D)   -10 degrees Celsius
(E)   -40 degrees Celsius

11.   If $1/A < 1/B$, then which of the following must be true:

(A) $A > B$
(B) $A < B$
(C) $A = B$
(D) $(A + 1) < B$
(E) $(A + B) = 0$

12.   An equilateral triangle with side length *x* has the same perimeter as a regular octagon with side length 12. What is *x*?

(A)   4
(B)   12
(C)   24
(D)   32
(E)   36

13.   How many triangles are contained within a regular shaped star (see image below)?

(A)   5
(B)   8
(C)   9
(D)   10
(E)   12

14.   Out of every minute, a traffic light is red for 25 seconds, green for 35 seconds, and yellow for 5 seconds. Over the course of a day, how long is the traffic light lit up red?

(A)   2 hours
(B)   6 hours
(C)   10 hours
(D)   12 hours
(E)   14 hours

15.   Multiply $(x + 3y)(2x – 6) =$

(A) $2x^2 + 6x + 6xy – 18y$
(B) $2x^2 – 6x + 6xy + 18y$
(C) $2x^2 + 18y$
(D) $2x^2 – 18y$
(E) $2x^2 – 6x + 6xy – 18y$

16.   The probability that Jay will go to the movies tonight is 2/3. The probability that Gene will go to the movies tonight is 1/5. What is the probability that neither Jay nor Gene go to the movies tonight?

(A)   2/15
(B)   4/15
(C)   8/15
(D)   2/3
(E)   1/5

17.   A 2-foot by 3-foot by 5-foot tank completely filled with water is poured into a 4-foot by 4-foot by 4-foot cubical tank.   When the water is poured into the new tank, how much of the new tank is filled with water?

(A)   46.875%
(B)   48.675%
(C)   50%
(D)   50.275%
(E)   53.725%

18.   Consider the two parallel lines $y = x + 2$ and $y = x – 2$. What is the shortest (perpendicular) distance between these two lines?

(A)   2 units
(B)   4 units
(C) $2\sqrt{2}$ units
(D) $4\sqrt{2}$ units
(E) $2\sqrt{3}$ units

19. What is the area of a triangle graphed in the coordinate plane having vertices (1, 3), (-4, 5) and (-2, 7)?

(A)   6 square units
(B)   6.5 square units
(C)   7 square units
(D)   7.5 square units
(E)   8 square units

20. If $(x + y) = 4$ and $(x - y) = 7$, what does the product xy equal?

(A)   -16.5
(B)   -8.25
(C)   8.25
(D)   16.5
(E)   It cannot be determined from the information given

21. What is the sum of the first 46 positive even integers?

(A)   2024
(B)   2068
(C)   2070
(D)   2114
(E)   2162

22. What is the sum of all possible solutions x to the equation $\frac{2}{x} = \frac{x}{2}$

(A)   -2
(B)   -1
(C)   0
(D)   1
(E)   2

23. Suppose that $\sqrt{a + b} = \sqrt{a} + \sqrt{b}$ for nonnegative numbers a and b. Which of the following must be true?

(A)   Neither a nor b can be 0.
(B)   Both a and b must be 0
(C) a must be 0
(D) b must be 0
(E)   Either a or b must be 0

24.   What is the sum of the digits of 11^4?

(A)   4
(B)   8
(C)   12
(D)   16
(E)   20

25.   What is the distance on the number line between the numbers 11 and -5?

(A)   5
(B)   6
(C)   11
(D)   16
(E)   21

STOP
This is the end of the test.

Test #5, Section 1

30 Minutes, No Calculator

Following each problem in this section, there are five suggested answers. Work each problem out, look at the five suggested answers, and decide which one is best.

1. Jessie likes to sell brownies. If she sells 2-inch by 3-inch brownie squares for $1.15 each, how much will she make from baking an 8-inch by 12-inch pan of brownies?

(A) $1.15
(B) $4.60
(C) $9.20
(D) $18.40
(E) $20.70

2. $4 \times 14 \times 20 = 5 \times 8 \times$ ___?

(A) 7
(B) 14
(C) 20
(D) 24
(E) 28

3. Marta goes to the store and buys a cake for $4.15, milk for $3.42, and a candy bar for $1.19. An 8.5% sales tax is added to these items. To the nearest cent, what is Marta's final bill?

(A) $8.76
(B) $9.13
(C) $9.50
(D) $9.51
(E) $9.53

4.	A room is filled with motorcycles and cars. In total, there are 32 vehicles and 104 wheels. What is the appropriate system of equations that one can use to solve this?

(A)	$x + y = 32$
	$2x + y = 104$

(B)	$x + y = 104$
	$2x + 4y = 32$

(C)	$x + 4y = 104$
	$2x + y = 32$

(D)	$2x + y = 32$
	$4x + 2y = 104$

(E)	$x + y = 32$
	$2x + 4y = 104$

5.	How many solutions (x, y) are there to $(x + 2)(y - 3) = 28$, where x and y are integers?

(A)	6
(B)	10
(C)	11
(D)	12
(E)	14

6.	Marla lives 50 miles north of the airport, Lori lives 50 miles south of the airport, and there is a straight road that directly connects Lori and Marla's homes. Lori and Marla both leave their homes to visit each other at exactly 9AM. Lori drives at 30 miles per hour and Marla at 20 miles per hour. Where will the two women meet each other?

(A)	10 miles north of the airport
(B)	5 miles north of the airport
(C)	2 miles north of the airport
(D)	2 miles south of the airport
(E)	5 miles south of the airport

7.   If $x^3 < x < x^2$, then which of the following must be true?

(A) $x > 1$
(B) $0 < x < 1$
(C) $-1 < x < 0$
(D) $x < -1$
(E)   This is true for no value of x.

8.   What is the surface area of a $3 \times 5 \times 7$ rectangular prism?

(A)   90 square units
(B)   105 square units
(C)   126 square units
(D)   142 square units
(E)   210 square units

9.   At noon, the minute and hour hand of a clock are at the same location. To the nearest second, when will be the next time that the hands of the clock will be exactly atop each other?

(A)   1:05 P.M.
(B)   15 seconds after 1;05 P.M.
(C)   27 seconds after 1;05 P.M.
(D)   43 seconds after 1:05 P.M.
(E)   1:06 P.M.

10.   15% of $2x$ is the same as what percent of $5x$?

(A)   4%
(B)   6%
(C)   8%
(D)   9%
(E)   10%

11.   If $x/y = 2/3, y/z = 4/7$, and $z = 5$, what does x equal?

(A)   40/21
(B)   10/3
(C)   20/7
(D)   10/21
(E)   20/21

12. Bernie has 40 inches of wire that he wishes to bend into a rectangular shape. What is the largest possible area he can enclose with this wire?

(A)   19 square inches
(B)   50 square inches
(C)   80 square inches
(D)   100 square inches
(E)   120 square inches

13. $\dfrac{4532}{7} =$

(A)   4000/7 + 300/7 + 50/7 + 2/7
(B)   4000/7 + 500/7 + 30/7 + 2/7
(C)   45/7 + 32/7
(D)   4000/7 + 53/7 + 2/7
(E)   4000/7 + 200/7 +50/7 + 3/7

14. Under what circumstances does $(x + y)^2 = x^2 + y^2$?

(A) x must equal 0
(B) y must equal 0
(C) x and y must equal 0
(D) x or y must equal 0
(E) $(x - y)$ must equal 0

15. Consider the equation $y = 3x + b$. If the x-intercept of this equation is 7, what must be the value of b?

(A)   -21
(B)   -7
(C)   0
(D)   7
(E)   21

16. The product of 0.4134 and 599.871 is closest to:

(A)   2.4
(B)   24
(C)   48
(D)   240
(E)   480

17.	What is the sum: 203 + 204 + 205 + ... + 461 + 462?

(A)	85326
(B)	85988
(C)	86190
(D)	86450
(E)	86652

18.	Solve for x: $\frac{4}{5}x + 1\frac{2}{7} = \frac{9}{11}$

(A)	$x = 45/77$
(B)	$x = -45/77$
(C)	$x = 35/55$
(D)	$x = -35/55$
(E)	$x = 11/35$

19.	Four triangles each with an area of 36 combined have the same area as a square with side length of what?

(A)	2
(B)	6
(C)	10
(D)	12
(E)	14

20.	A triangle with angle measures of 31°, 67°, and 82° can be classified as which of the following:

	I.	scalene triangle
	II.	acute triangle
	III.	isosceles triangle

(A)	I only
(B)	II only
(C)	III only
(D)	I and II
(E)	I and III

21.   The average height of girls in a class is 62 inches. The average height of boys in the same class is 71 inches. If there are twice as many boys as girls, what is the average height of all people in the class?

(A) $67\frac{1}{3}$ inches

(B) $67\frac{2}{3}$ inches

(C)   68 inches

(D) $68\frac{1}{3}$ inches

(E) $68\frac{2}{3}$ inches

22.   45% of 70 equals 2/3 of what number?

(A)   46.75
(B)   47.25
(C)   47.75
(D)   48.25
(E)   49.25

23.   Shonté and her brother Fred are playing a game of basketball. Shonté tells her brother that if he makes a shot and she misses it, then she'll give him $50. The probability of Fred making his shot is 2/3, and the probability Shonté makes it is 2/5. What is the probability that Fred will be getting $50?

(A)   1/5
(B)   1/3
(C)   2/15
(D)   2/5
(E)   3/5

24.   If $6x - 7 = 3x + 1$, solve for x:

(A) $x = -8/3$
(B) $x = -3/8$
(C) $x = 0$
(D) $x = 3/8$
(E) $x = 8/3$

25.   If $35 \times N = 35$, what is $35 - N$?

(A)   0
(B)   1
(C)   34
(D)   35
(E)   36

STOP
This is the end of the test.

Test #5, Section 2

30 Minutes, No Calculator

Following each problem in this section, there are five suggested answers. Work each problem out, look at the five suggested answers, and decide which one is best.

1. Consider the number $1 \times 2 \times 3 \times 4 \times 5 \times 6 \times 7 \times 8 \times 9 \times 10 \times 11 \times 12 \times 13$. Which of the following numbers does not divide this number?

(A) 88
(B) 132
(C) 153
(D) 198
(E) 231

2. Leap years (eg, an extra day, February 29, is added to the calendar) occur on every year that is divisible by 4 <u>except</u> for years whose number is divisible by 100 and not 400. Between 1887 and 2014, how many leap years were there?

(A) 29
(B) 30
(C) 31
(D) 32
(E) 33

3. If 2 cups of sugar are needed to bake 12 cookies, how many cups of sugar are needed to bake 100 cookies?

(A) $16\frac{2}{3}$ cups

(B) 17 cups

(C) $17\frac{1}{3}$ cups

(D) $17\frac{2}{3}$ cups

(E) 18 cups

4.   How many different ways can 5 people be ordered in a line?

(A)   5
(B)   20
(C)   60
(D)   90
(E)   120

5.   Which of the following is true about three consecutive integers:

I.   Their product is always divisible by 6
II.   Their sum is always even
III.   The largest divided by the smallest is always less than 4

(A)   I only
(B)   II only
(C)   III only
(D)   I and III
(E)   II and III

6.   A 24-foot tall flagpole casts a 7-foot shadow on the ground. How far is the straight-line distance from the tip of the shadow to the top of the flagpole?

(A)   7 feet
(B)   17 feet
(C)   24 feet
(D)   25 feet
(E)   31 feet

7.   How many positive integers less than 150 have exactly 3 divisors?

(A)   9
(B)   10
(C)   11
(D)   12
(E)   13

8.   Yannick drops a ball from his balcony, which is 10 feet above the ground. The ball hits the ground and bounces back to a height of 6 feet. In fact, after every successive bounce, the ball will return to a height of 60% of its previous height. How high does the ball bounce after its third bounce?

(A)   10 feet
(B)   6 feet
(C)   3.6 feet
(D)   2.16 feet
(E)   1.296 feet

9.   Van tosses a fair coin 500 times. What is the probability that she will get a head on the 500th toss?

(A) $(1/2)^{500}$
(B)   1/500
(C)   1/250
(D)   1/2
(E)   499/500

10.   Julia is given a piece of wire that is 30 inches long. She chooses to bend it into the shape that will give it the greatest area. Which shape will she bend it into?

(A)   equilateral triangle
(B)   square
(C)   regular pentagon
(D)   regular hexagon
(E)   circle

11.   Often license plates on cars take the form *NNN LLLL*, where *N* is a number between 0 and 9 and *L* is a letter of the alphabet. How many different combinations can me made to identify a license plate?

(A) $10^3 \times 26^4$
(B) $26^3 \times 10^4$
(C) $26 \times 3 \times 10 \times 4$
(D) $26^3 \times 10 \times 4$
(E) $26 \times 3 \times 10^4$

12. Please express $6\frac{3}{4}\%$ as a decimal:

(A) 6.75
(B) 0.675
(C) 0.0675
(D) 0.00675
(E) 0.000675

13. Which of the following lines is perpendicular to $6x - 9y = 11$:

(A) $6x + 8y = 2$
(B) $6x + 9y = -5$
(C) $3x + 2y = 4$
(D) $12x - 18y = -7$
(E) $-9x - 10y = 2$

14. What is the greatest common factor of 24, 42, and 72?

(A) 2
(B) 3
(C) 4
(D) 6
(E) 8

15. An elevator is on the fifteenth floor of a building. If it goes up two floors, then down seven floors, then up five floors, then down four floors, what floor will the elevator now be on?

(A) 9
(B) 10
(C) 11
(D) 15
(E) 17

16. Alex and Brad begin running clockwise around a circular track at the same time. The track has a circumference of 600 meters. Alex runs at a speed of 6 kilometers per hour and Brad runs at a speed of 4 kilometers per hour. After how long will Alex pass Brad on the track? (Note: 1 kilometer = 1000 meters)

(A)   10 minutes
(B)   12 minutes
(C)   18 minutes
(D)   20 minutes
(E)   30 minutes

17. Evaluate: $2 + \dfrac{1}{2 + \frac{1}{2}}$

(A)   2
(B)   2.25
(C)   2.4
(D)   2.5
(E)   2.6

18. Solve for x: $2.5x - 1 = 8$

(A)   3.2
(B)   3.3
(C)   3.4
(D)   3.5
(E)   3.6

19. A 10-gallon tank currently is half full and filled with 20% water and 80% oil. How much water do I need to add to make the tank be 40% filled with water?

(A)   1 gallon
(B)   1⅓ gallons
(C)   1½ gallons
(D)   1⅔ gallons
(E)   2 gallons

Questions 20 and 21 refer to the pie chart below:

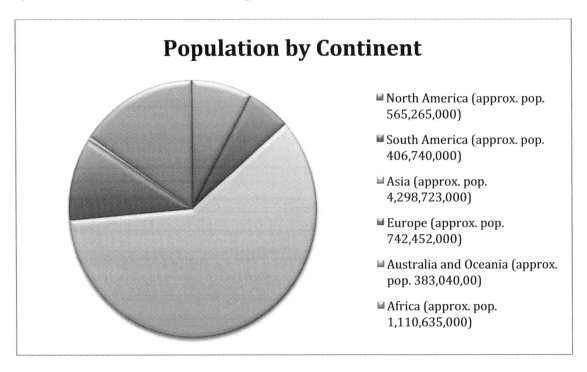

Population by Continent

- North America (approx. pop. 565,265,000)
- South America (approx. pop. 406,740,000)
- Asia (approx. pop. 4,298,723,000)
- Europe (approx. pop. 742,452,000)
- Australia and Oceania (approx. pop. 383,040,00)
- Africa (approx. pop. 1,110,635,000)

20.   To the nearest percent according to the chart, the population of Asia represents what percent of the world's population?

(A)   55%
(B)   60%
(C)   65%
(D)   70%
(E)   75%

21.   The population of North America represents what percentage of the population of the Americas?

(A)   56%
(B)   57%
(C)   58%
(D)   59%
(E)   60%

22.   If I take the equation $y = (1/2)x + 5$ and reflect it over the y-axis, what will be the equation of the new line formed?

(A) $y = 2x + 5$
(B) $y = -2x - 5$
(C) $y = -(1/2)x - 5$
(D) $y = -(1/2)x + 5$
(E) $y = -2x + 5$

23.   A map has a scale such that one inch represents 80 miles. How far apart on this map are two cities that are located 2000 miles apart in reality?

(A)   18 inches
(B)   20 inches
(C)   22 inches
(D)   25 inches
(E)   28 inches

24. $(5.6 \times 10^7) \times (2.3 \times 10^4) =$

(A) 1.288×10^{10}
(B) 1.288×10^{11}
(C) 1.288×10^{12}
(D) 12.88×10^{10}
(E) 12.88×10^{12}

25. $(1/2)^2 + (1/3)^3 + (1/4)^4 =$

(A)   35/216
(B)   2011/6912
(C)   1417/2304
(D)   451/678
(E)   643/768

STOP
This is the end of the test.

ANSWER KEY

Test #1, Section 1:

1.	E
2.	B
3.	A
4.	D
5.	B
6.	A
7.	D
8.	D
9.	B
10.	B
11.	D
12.	E
13.	B
14.	D
15.	E
16.	A
17.	B
18.	E
19.	E
20.	D
21.	C
22.	B
23.	A
24.	C
25.	B

Test #1, Section 2:

1.	A
2.	A
3.	D
4.	E
5.	A
6.	B
7.	D
8.	C
9.	B
10.	A
11.	D

12.	B
13.	A
14.	B
15.	D
16.	A
17.	D
18.	D
19.	C
20.	D
21.	B
22.	C
23.	E
24.	C
25.	E

Test #2, Section 1:

1.	B
2.	C
3.	D
4.	D
5.	A
6.	D
7.	D
8.	A
9.	C
10.	D
11.	B
12.	B
13.	C
14.	E
15.	E
16.	B
17.	E
18.	E
19.	C
20.	B
21.	E
22.	C
23.	A
24.	D
25.	D

Test #2, Section 2:

1. C
2. D
3. C
4. A
5. C
6. E
7. B
8. B
9. A
10. B
11. C
12. B
13. B
14. A
15. E
16. C
17. D
18. A
19. E
20. A
21. E
22. B
23. E
24. B
25. E

Test #3, Section 1:

1. B
2. A
3. D
4. E
5. A
6. D
7. B
8. A
9. D
10. D
11. E
12. D
13. C
14. A
15. C

16. B
17. A
18. B
19. A
20. D
21. E
22. C
23. A
24. D
25. B

Test #3, Section 2:

1. A
2. C
3. D
4. C
5. A
6. E
7. E
8. D
9. C
10. D
11. C
12. D
13. E
14. B
15. E
16. A
17. C
18. C
19. B
20. A
21. B
22. C
23. D
24. B
25. B

Test #4, Section 1:

1. A
2. C
3. C
4. D
5. D
6. C
7. C
8. B
9. C
10. B
11. C
12. D
13. C
14. C
15. E
16. B
17. C
18. B
19. D
20. C
21. B
22. C
23. A
24. B
25. D

Test #4, Section 2:

1. B
2. A
3. D
4. B
5. D
6. E
7. D
8. C
9. E
10. E
11. A
12. D
13. D
14. C

15. E
16. B
17. A
18. C
19. C
20. B
21. E
22. C
23. E
24. D
25. D

Test #5, Section 1:

1. D
2. E
3. C
4. E
5. D
6. A
7. D
8. D
9. C
10. B
11. A
12. D
13. B
14. D
15. A
16. D
17. D
18. B
19. D
20. D
21. C
22. B
23. D
24. E
25. C

Test #5, Section 2:

1. C
2. C
3. A
4. E
5. A
6. D
7. D
8. D
9. D
10. E
11. A
12. C
13. C
14. D
15. C
16. C
17. C
18. E
19. D
20. B
21. C
22. D
23. D
24. C
25. B

ABOUT THE AUTHOR

Justin Grosslight is an academic entrepreneur interested in examining relationships between science and business. He is especially intrigued by how networks operate (quantitatively and qualitatively), both from historical and from contemporary perspectives. He holds degrees in history and mathematics from Stanford, a history of science degree from Harvard, and has published in all three fields. He is passionate about business, entertainment, academia, and writing, and enjoys helping talented youth thrive in their intellectual pursuits.

Justin has had years of experience in training students for their SAT®, SSAT®, ACT®, GMAT®, GRE®, AP® Calculus, IB® mathematics, SAT® physics subject test, and SAT® mathematics subject tests. He is a national merit scholar who received a perfect 800 on his SAT® math exam, perfect 800 on his SAT® math level 2 subject test, perfect 170 on his GRE® math exam, perfect 5 on his AP® Calculus BC exam (as a sophomore), perfect 5 on his AP® US History Exam, perfect 6 on his GMAT® writing exam, a perfect 8 on his GMAT® integrated reasoning exam, and near perfect scores on all of his other exams. He has published widely in both the humanities and in mathematics.

Several of Justin's students have received perfect scores on sections of their SAT® and ACT® exams, on their SAT® math subject test, and on their SAT® physics subject test. They have gained admission into prestigious universities such as Stanford University, NYU, UCLA, The University of Pennsylvania, and Oberlin College.

ABOUT MANDA EDUCATION

Manda Education is a test preparation program that believes in helping students whet their educational skills through personalized training. Realizing that different students have different needs, Manda Education believes that the best way to train students was to mentor them in both a personalized and intensive framework. Our aims are threefold:

(1)   To help students achieve test scores that will gain them admittance to universities of their choice
(2)   To develop writing, communication, quantitative, and analytical skills that will help students flourish in a global context
(3)   To instill values of character and responsibility in students that will help students succeed in their personal and professional endeavors

We also believe that many students and professionals can benefit from our books, so we have released them for sale to the public.

Made in the USA
San Bernardino, CA
21 July 2018